The Eurasian Beaver

Róisín Campbell-Palmer, Derek Gow, Robert Needham,
Simon Jones and Frank Rosell

Illustrations by Rachael Campbell-Palmer

Pelagic Publishing | www.pelagicpublishing.com

Published by Pelagic Publishing
www.pelagicpublishing.com
PO Box 725, Exeter EX1 9QU, UK

On behalf of The Mammal Society
www.mammal.org.uk
3 The Carronades, New Rd, Southampton SO14 0AA

The Eurasian Beaver
ISBN 978-1-78427-034-6 (Pbk)
ISBN 978-1-78427-035-3 (ePub)
ISBN 978-1-78427-036-0 (Mobi)
ISBN 978-1-78427-040-7 (PDF)

This book should be cited as:
Campbell-Palmer, R., Gow, D., Needham, R., Jones, S. and Rosell, F.
(2015) *The Eurasian Beaver*. Exeter: Pelagic Publishing

British Library Cataloguing in Publication Data
A catalogue record for this book is available from the British Library.

Cover image
Eurasian beaver, *Castor fiber*, at the Devon Beaver Project, UK
Copyright © 2012 David Plummer www.davidplummerimages.co.uk

Contents

About the authors

Róisín Campbell-Palmer is currently the Conservation Projects Manager for the Royal Zoological Society of Scotland, where she has worked for 12 years, beginning as an animal keeper. For the last five years she has focused on the reintroduction of beavers to Scotland through her role as the Field Operations Manager for the Scottish Beaver Trial. She is also undertaking her PhD in beaver health and welfare through Telemark University College (TUC), Norway, publishing a number of journal articles on this subject. Most recently Róisín has completed a trapping for health and genetic screening project on Tayside beavers for Scottish Natural Heritage. She also sits on the Beaver Advisory Committee for England and the Scottish National Species Reintroduction Forum. Róisín completed her honours degree in Zoology at the University of Glasgow and her MSc in Applied Animal Behaviour and Welfare at the University of Edinburgh. She is passionate about native wildlife conservation.

Derek Gow is a freelance ecologist who has worked with beavers and water voles in Britain for over 20 years. During this time he has organised the quarantine and import of beavers from Poland, Germany and Norway. He has travelled widely in Europe to study beavers in a range of different landscapes. Derek has advised both Natural England and Countryside Council for Wales's feasibility studies regarding beaver re-introduction.

Robert Needham has always been passionate about wildlife and its conservation, and more recently reintroductions of former native species. His studies have included a BSc in Wildlife Management and an MSc in Ecology. He has worked in some spectacular landscapes in

Norway and Scotland and been privileged to work with some of Europe's most charismatic species including Eurasian Lynx, Wolverine, Capercaillie, and last but not least the Eurasian beaver. He has spent the past three and a half years working on the Scottish Beaver Trial and considers it an honour to be involved in the first trial reintroduction of a mammal species in the UK. He is currently undertaking a PhD at the University of Southampton researching the relationship between beaver dams and the movement of fish.

 Simon Jones has recently been appointed as Director of Conservation at the Scottish Wildlife Trust, and was the Project Manager for the Scottish Beaver Trial between 2008 and 2014. Simon's career in wildlife conservation and reserve management extends over 20 years and across various parts of the UK and abroad. He lives in rural Stirlingshire with his family and four bikes.

 **Frank Rosell** is a renowned beaver ecologist and is presently a professor in behaviour ecology at Telemark University College in Norway. He completed his BSc degree in Chemical Engineering in 1991 at Østfold University College. He finished his MSc thesis in 1994 and his PhD in 2002, both at Norwegian University of Science and Technology. He has worked at TUC since 1994 and has taught a number of different courses in wildlife management, conservation biology and behavioural ecology. He has a broad scientific interest and has published more than 80 scientific papers with peer review. At TUC he is a member of the research committee and the PhD committee in Ecology. He was also a part of advisory board of the Scottish Beaver Trial.

Overleaf: Adult female beaver feeding on Loch Buic, Scottish Beaver Trial
(Philip Price www.lochvisions.co.uk)

CHAPTER 1

Introduction

Amongst conservation bodies, various land-use organisations and the wider public there is a growing interest in the reintroduction of the formerly native Eurasian beaver (*Castor fiber*) to Britain. The pros and cons of restoring this species have been well aired in the popular media for well over ten years. A five year long trial reintroduction project – the Scottish Beaver Trial – was granted a licence by the Scottish Government in 2008. Discussions on licensed beaver reintroductions in England and Wales also gathered pace during this period and feasibility studies by Natural England and the Countryside Council for Wales were completed to advise this process. Despite no widespread formal reintroduction programmes taking place in Britain to date, field sign evidence exists, suggesting the presence of un-licensed free-living beavers in various parts of mainland Britain.

The aim of this booklet is to raise awareness of the Eurasian beaver as a former British native mammal and to provide factual information regarding its biology, behaviour and ecology. It can be used as a guide to enable the identification of field signs and to provide information on the design of beaver monitoring programmes or survey work. Although beavers are not currently extensively found in Britain the increased likelihood of future reintroductions suggests that an understanding of their ecology and management will be an essential long term component of their successful coexistence in human dominated landscapes.

Beaver folklore and history in Britain

Archaeological evidence of beavers, such as preserved gnawed timber and bones, has been found in a number of sites throughout England, Wales and Scotland. Along with related place names such as Beverley, Beverstone and Bevercots, carvings and historical references testify to their former abundance throughout Britain (Coles 2006).

In 1188 the travelling monk, Giraldus Cambrensis, provided one of the earliest and most fanciful descriptions of British beavers: 'The beavers, in order to construct their castles in the middle of rivers, make use of the animals of their own species instead of carts, who,

1

by a wonderful mode of carriage, convey the timber from the woods to the rivers. Some of them, obeying the dictates of nature, receive on their bellies the logs of wood cut off by their associates, which they hold tight with their feet, and thus with transverse pieces placed in their mouths, are drawn along backwards, with their cargo, by other beavers, who fasten themselves with their teeth to the raft' (Cambrensis 1188). In 1526 Hector Boece recorded beavers as being abundant around Loch Ness. By 1577 William Harrison, the Canon of Windsor, describes the beaver in shape '… as the bodie vnto a monsterous rat: the beast also it selfe is of such force in the teeth, that it will gnaw an hole through a thicke planke, or shere thorough a dubble billet in a night; it loueth also the stillest riuers' (Coles 2006).

In Britain the 1566 'Act for the Preservation of Grayne' saw a system of bounty payments made by Parish Constables for the heads of specific wild mammals and birds considered to be vermin. These records provide evidence that beavers could have survived in small numbers in some parts of Britain until more recent times. At Bolton Percy, a village near York, the Church warden's account of 1789 records two pence being paid to John Swail for 'a bever head'. Bolton Percy is connected by the rivers Wharfe and Washburn to sites near Harrogate where the place names Beaver Dyke and Beaver Hole still remain. In 1904 a Yorkshire naturalist called Edgar Bogg reported that he had been told by an elderly man that his grandfather recalled beavers living in an area known as Oak Beck in his youth. Bogg dated this time to around 1750. Oak Beck is only separated by 50 km of watercourse from Bolton Percy and this distance is well within the exploration range of sub-adult beavers seeking to establish independent territories (Coles 2006).

The Catholic Church considered beavers to be primarily aquatic and therefore categorised as fish, as opposed to 'creatures which creapeth upon the land' (Leviticus 11, King James Bible). This distinction meant that their flesh, like other designated species such as seals and otters, could be consumed during fasts and holidays, when the eating of 'meat' was forbidden. It is likely that persecution through hunting could have played a significant role in the decline of the beaver as hunting during the Lenten fast of 40 days at Easter would have entailed killing adult female beavers when they were either heavily pregnant or suckling young.

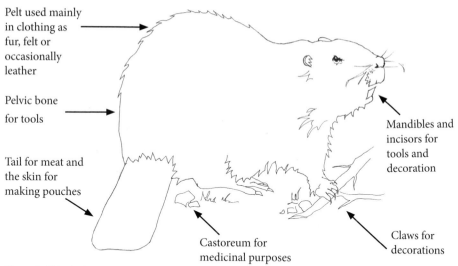

Pelt used mainly in clothing as fur, felt or occasionally leather

Pelvic bone for tools

Tail for meat and the skin for making pouches

Mandibles and incisors for tools and decoration

Castoreum for medicinal purposes

Claws for decorations

Figure 1 Historical uses of beaver (adapted from Coles 2006).

Beaver castoreum, a viscous liquid produced by two castor sacs positioned inside their cloaca (a single opening for reproductive, urinary and intestinal tracts), was also a highly prized commodity. It is used by beavers for chemical communication, including scent marking their territories. Amongst other components it contains a concentrate of salicylic acid, which is derived from the willow bark they consume. Salicylic acid is an effective ingredient of the modern drug aspirin and castoreum has known analgesic and anti-inflammatory properties. Although principally used as an agent of pain relief in medieval times it was also applied to a wider range of ailments such as constipation, hysteria, dysmenorrhea, abortion, toothache and ulcers, while beaver skin was applied as an aid to heal gout and rheumatism (Pilleri 1986). Additionally castoreum has been widely used in perfumery, where it provides natural musk tones to some well-known scents.

In the past the soft, luxuriant under fur of the beaver was expensive and in great demand. In historical times it was used for cradle blankets and a harp bag uncovered in the excavation of the Viking burial ship at Sutton Hoo, Suffolk was made from the in-turned skin of at least two adult beavers which had been sewn together (Bryony Coles personal communication). In more recent times beaver fur was shaved from the skins and pressed into sheets of felt, which were coloured and moulded into fashionable hats that were both waterproof and durable. With care a beaver hat could easily last a lifetime and organisations like the British Army and Royal Navy provided a virtually insatiable market for this product. A British Act of Parliament in 1638 made it illegal for hats to be made from anything other than beaver fur ('beaver stuff and beaver wool'). This resulted in vast numbers of Eurasian and North American beavers (*C. canadensis*) being killed and imported for this purpose.

By the fifteenth century the trade in beaver furs in Scotland was no longer economically viable due to over-exploitation and while oral tradition suggests they may have survived in and around the Loch Ness and Lochaber areas until the late seventeenth century, there is no further mention of their presence after this time (Coles 2006). In Wales memories of their former presence were recalled in the Ogwen Valley until the end of the eighteenth century, where they are thought to have gone extinct as a species by the early 1700s (Coles 2006). There is no evidence for their existence in Ireland.

CHAPTER 2

Beaver Biology
and Behaviours

Taxonomy

The Eurasian and the North American beaver are the only surviving members of the once large family of Castoridae. Beavers are the second largest species of rodent in the world (the largest being the capybara, *Hydrochoerus hydrochaeris*, from South America), although their extinct relative the North American giant beaver (*Castoroides leiseyorum*) was as big as a modern bear. This species measured up to 2.5 m in length and weighed up to 200 kg (Miller *et al.* 2000). Another extinct beaver species, *Palaeocastor magnus*, excavated deep, spiral burrow systems in arid grasslands in order to trap water. Several thousand of these unusual structures that can descend to a depth of nearly 3 m and are called 'Devils Corkscrews' have now been identified in North America (Martin 1994).

Both modern beaver species, Eurasian and North American, are physically very similar making them hard to distinguish in the field. They have comparable ecological requirements and behavioural patterns and were once considered to be a single species. Chromosome analysis has identified that they are quite distinct (Eurasian beavers have 48 pairs of chromosomes whilst North American beavers have 40 pairs, Lavrov and Orlov 1973) and the two species will not interbreed even in captivity to produce viable offspring. It is now believed that the two species diverged about 7.5 million years ago when beavers first colonised North America from Eurasia across the Bering Strait land bridge (Horn *et al.* 2011).

Identification

Beavers are large rodents (adults >20 kg) with paddle-like tails covered with visible scales (**Figure 2**). They have broad heads with powerful facial muscles and large, orange, incisor teeth, which enable them to fell trees and process wood with ease (**Figure 3**).

Beavers do not attain their full adult size until around three years of age. Adults of both sexes have similar head and body lengths although females are on average ~1 kg heavier than males (Wilsson 1971). Mature

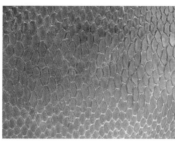
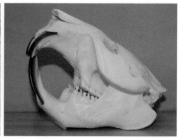

Figure 2 Beaver tail scales with very sparse and short hairs dispersed between them (S. Jones).

Figure 3 Beaver skull with prominent large orange-fronted incisors (R. J. Needham).

beavers have a head and body length of around ~80 cm and a tail length of ~30 cm. The sexes are impossible to distinguish through visual observations unless a female is pregnant or lactating. At this time prominent nipples on her chest may be visible between her forelegs (**Figure 4**).

The coat colour of beavers ranges from light golden-brown through to black. Although white or fawn beavers have occasionally been recorded in both the wild and captivity, these are rare. Some North American beavers have light tan cheeks, flanks and belly patterns. Though beavers have small ears, their hearing is good and they have an excellent sense of smell. They have small eyes and poor vision, so mainly react to unfamiliar movements.

Figure 4 Visible nipples of a lactating female (J. McIntyre).

Adaptations for a semi-aquatic lifestyle

Beavers spend a large amount of their active time in the water foraging and maintaining their territories. They are very able swimmers and can easily move rapidly in the water for short distances when they feel threatened. Their ears, eyes and nose are levelly positioned on the top of their heads to ensure that they have full use of their senses while swimming (**Figure 5**).

Beavers breathe through their nose and can constrict their nostrils to prevent water entry when diving. They have internal ear flaps that they can fold and extra internal hairs to trap air and reduce water entry. Unlike many other mammals their epiglottis is located inside the nasal cavity rather than in the back of the throat and this prevents water from entering their larynx and trachea. There is also a raised section at the back of the tongue to ensure a watertight fit. This adaptation, along with the ability to close their lips behind their incisors, means they are able to forage underwater, cut branches or pull up plant roots without swallowing water.

Large webbed hind feet (**Figure 6**) with powerful hind leg muscles provide most of the thrust they require for swimming. They tuck their forefeet up into their chest when they move through the water and generally use their tail as a rudder, though it can be moved in synchrony with thrusts from their hind-legs to enable swimming at speed (Wilsson 1971).

Figure 5 Ears, eyes and nose are all set high enabling all senses to be employed whilst swimming (S. Gardner).

Figure 6 Large webbed hind feet are perfectly adapted for swimming (R. J. Needham).

Beaver fur consists of two different hair types – long, coarse guard hairs and a dense, soft under fur. This double structure traps a layer of air next to the skin to provide insulation against the cold and helps repel water. Beavers are meticulous in the maintenance of their fur and spend a lot of time grooming. They have a specifically adapted claw (split nail, **Figure 7**) on the second toe of each hind foot, which is formed by one nail growing immediately alongside the other, with only a tiny space in-between. This functions as an effective comb for removing parasites and keeping hairs separate.

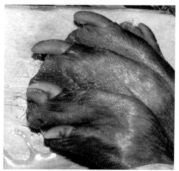

Figure 7 Grooming claw, second toe on hind feet (R. Campbell-Palmer).

Although beavers can remain submerged for up to 15 minutes this behaviour is unusual and generally occurs when they feel threatened. Their adaptations for diving include an ability to reduce their heart rate (bradycardia) to around half its normal function coupled with an increase in blood flow to the brain. This is achieved by reducing the supply to all their other organs with the exception of their heart, lungs and adrenal glands (McKean 1982). Beavers are able to tolerate high concentrations of carbon dioxide in their body tissues (Irving 1937).

Adaptations for a life on land

When walking on land beavers are generally slow and cumbersome although they can run fast over short distances. When threatened they usually tend to head for the nearest source of water. While they normally walk on all fours they can transport building materials over short distances by clasping it between their front paws and their chest and waddling on their hind legs for short distances, although regular stops are required. When they reach their destination they drop their materials and either weave or push them into position at the lodge or dam. They will also rise up on their hind legs to fell trees or gnaw branches. Whilst their fore limbs and paws are relatively small they are dexterous and strong (**Figure 8 and 9**). Beavers use their forepaws for digging, holding

Figure 8 Dexterous forepaws (R. J. Needham).

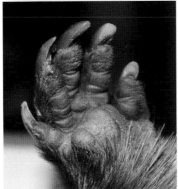

Figure 9 Dexterous forepaws (R. J. Needham).

food, carrying and manipulating earth or vegetation using their smallest digit or 'little finger' in a similar way to an opposable thumb. Only juvenile beavers use their forelimbs for swimming.

Beaver tails are covered in skin scales interspersed with sparse, short, stubby hairs (**Figure 2**). Internally they mainly consist of fatty tissue, which acts as an important energy reserve. Beavers use their tails as a balancing aid when they rise up on their hind legs to gnaw tree trunks or branches or carry building materials. During territorial fights their tails are often bitten and this can result in unique patterning which can be used to identify an individual (see **Chapter 5, Figure 51**).

Breeding

In a beaver family only the adult male and female breed. Both sexes become sexually mature at around 20 months of age. Females come into oestrus between late December and February each year. Mating occurs in the water with the female floating while the male clings on to her flank (**Figure 10**). This process can last from 30 seconds to three minutes (Wilsson 1971).

After a gestation of 105–107 days, the female gives birth in the warm, dry, central chamber of a lodge (Wilsson 1971). Breeding pairs have one litter per year, usually two to four kits in total. The number of kits produced and their survival is affected by various factors such as the age of the parents, the surrounding population density of other beavers, the altitude of their location and habitat quality. Beaver kits are born fully furred and usually weigh between 300 to 700 grams. Their eyes are open within a few days of birth and although they will feed on their mother's milk for between two to three months, they can consume vegetation after their first week. The kits remain in the

Figure 10 Copulation in the water (replicated from R. Pollitt in Kitchener 2001).

lodge for the first one to two months of their lives (**Figure 11**) while their parents and older siblings bring leafy twigs and other vegetation for them to eat. After a period of practice diving and tentative exploration around their lodge, they rapidly start to independently forage further afield. Whilst they are small they can exhibit a rarely seen behaviour known as 'caravanning' where they swim in the wake of an older relative, often partially clinging on to their back 'hitching a lift'.

Social behaviours

Beavers live in family groups comprising of an adult breeding pair with their offspring from the current and the previous year's litter. Once paired, beavers tend to remain together until one of them is either displaced by another individual of the same sex or dies. The adult male and other family members all help to rear any kits. When they are approximately two years old they tend to disperse and establish their own territories.

Like most social mammals, interactions between individuals serve to reinforce family bonds (Wilsson 1971). Beavers spend a lot of time and energy grooming to maintain their fur quality. Grooming is an important bonding behaviour between adult beavers and their offspring. Kits spend a considerable amount of time play wrestling, feeding together or mock fighting over sticks. Social interactions are more common in younger animals and tend to decline with age. Beavers can produce a range of vocalisations and family members will commonly produce whining calls when they meet, with young animals tending to be much more vocal. Juveniles will communicate

Figure 11 Adult with ~two to three month old kit at Scottish Beaver Trial (S. Gardner).

with a range of mews, short, soft squeaking calls or repetitive crying noises to which the adults respond. These can be easily heard in the early evening when they are active outside a lodge or on occasion from within the lodge itself.

Wrestling bouts usually occur in the water when beavers grasp one another with their forepaws, thrust their noses together or push against each other. This behaviour is more common in younger siblings seeking to establish themselves. Aggressive interactions between family members are not common but can involve vocalisations and open mouthed lunges with their teeth exposed. These rarely result in actual bites. A subordinate individual will move away rather than confront a more dominant family member. However, savage encounters with intruders from other families are common. During these fights, deep, penetrating wounds are frequently inflicted by biting to the flanks, tails and rumps (**Figure 12**). Both individuals can be injured or scarred and severe injuries, which may become septic, can result in death.

Figure 12 Obvious scarring from beaver bites on the underside of a beaver pelt (R. J. Needham).

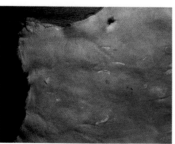

Defensive behaviours

Beavers are renowned for using their tails to generate alarm signals to warn other family members of approaching danger. 'Tail slapping', as the name suggests, involves a beaver raising its tail above the water, then bringing it down sharply to 'slap' the water's surface (**Figure 13**). This behaviour is usually followed by a dive. Tail slapping warns other family members of the presence of danger, frightens an attacker or lets a potential predator know that they have been seen. The presence of unfamiliar noises and odours (including castoreum from unknown beavers) can also stimulate tail slapping.

Figure 13 Tail slapping.

Beavers may 'freeze' when threatened; this behaviour can occur on land or while they are stationary underwater. This behaviour may help conceal the beaver and reduce the likelihood of its detection by a predator. Long dive times (up to 15 minutes) have been recorded for extremely threatened beavers, which have been seen pressing themselves against the bottom of the river or lake, where they remain motionless. Beavers can dive and swim considerable distances (~800 m) under water to escape danger. This is the most common reaction of disturbed animals (Wilsson 1971).

When beavers are threatened they can emit a very distinctive hissing growl, which may be accompanied with teeth grinding. If this warning is not sufficient they can then make mock charges towards a perceived source of danger and in extreme circumstances they will make contact and bite.

Chemical communication and scent marking behaviours

Chemical communication is highly developed in beavers, with scent marking regularly employed in territorial defence (Campbell-Palmer and Rosell 2010). Their principal sources of scent are obtained from the castoreum in their castor sacs and their anal gland secretions (AGS). These are located in the internal cavities between their pelvis and the base of their tail. The anal glands have separate openings, which are only visible when they are exposed during scent marking. Their secretions are coloured according to their sex: grey/white with a more paste-like consistency for females and yellowish brown and runny in consistency for males (this differs in North American beavers) (Rosell and Sun 1999), and can be manually extracted by experienced personnel (**Figure 14**).

Figure 14 Exposure of the anal papillae and AGS collection (Scottish Beaver Trial).

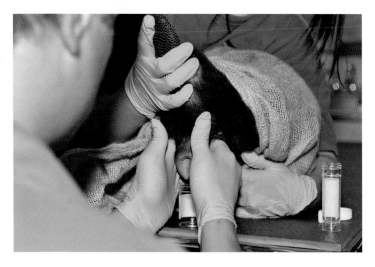

Beavers tend to scent mark on specially constructed 'mounds', which are often formed from gathered piles of mud or vegetation. These tend to be found along territorial boundaries or near feeding and resting sites close to the water's edge. Castoreum is mainly formed from dietary components and functions as the main scent used in territorial defence. Anal gland secretion is individually distinct and transmits information to other beavers on their sex, age, dominance status and relatedness. Families counter-mark and respond aggressively towards the scent of other beavers.

Food processing and foraging behaviour

In more recent times, thanks to stories such as *The Chronicles of Narnia* where a pair of beavers play key characters in the story, the beaver's partiality to fish has developed into an enduring myth. They are in fact completely herbivorous and will readily consume a wide range of bark, shoots and leaves of woody (predominantly broadleaf species), herbaceous and aquatic vegetation (**Figures 15 and 16**). During spring and summer 90% of their diet is comprised of herbaceous terrestrial, semi-emergent and aquatic species of plants (Nolet *et al.* 1995).

Figure 15 Wild beaver feeding on Japanese knotweed (*Fallopia japonica*), River Ericht, Scotland (P. Scott).

Figure 16 Wild beaver feeding on wild garlic (*Allium ursinum*), River Ericht, Scotland (P. Scott).

Beavers often dive repeatedly when they are feeding on submerged plants and although these feeding dives can last for between 5–6 minutes (Wilsson 1971) they are generally much shorter. Beavers do not hibernate and in the autumn and winter their diet tends to consist predominantly of tree bark and twigs. During the autumn they construct food caches outside their principal lodge or burrows by jamming branches into the underwater sediment immediately outside the lodge or burrow. A mass of lighter woody material is then interwoven through these mainstays. Beavers can access this material under ice in climates where winter conditions are severe and will readily gnaw diving holes through frozen water surfaces.

Although beavers can fell quite large trees (>1 m in diameter) they tend to favour smaller saplings (<5 cm diameter) in order to obtain their bark, side branches and leafy stems (Haarberg and Rosell 2006). Most feeding activity occurs at or close to the water's edge (usually within ~20 m) although the targeted felling of specifically desirable tree species such as poplars (*Poplus* spp.) can occur within a wider foraging range. When trees are felled their side branches or upper crowns are removed in manageable sections. These are then dragged back to the water to be consumed or incorporated into constructions such as lodges and dams. Beavers feed whole on very fine sticks and stalks by pushing them steadily into their mouth with their forepaws. They can fold leaves into small bundles and then eat these in the same manner. Beavers are creatures of habit and develop favoured sites on the water's edge known

Figure 17 Beaver rotating stick with its forepaws and removing bark with its incisors (note the use of the 'little finger').

as feeding stations where distinctive piles of peeled, woody debris accumulate. Although these tend to gradually disappear over time some have been identified in archaeological sites (Coles 2006).

Beavers remove the bark from thicker sticks with their incisors while rotating them with their forepaws (**Figure 17**). They strip the bark from felled tree trunks by placing their forepaws against the trunk and grasping the bark with their incisors, making upwards movements with their heads. Felled trunks can be cut into smaller, more manageable logs for transport and building. The distinctive noises of beavers feeding, particularly on woody or crisp green vegetation, is associated with a gnawing and rasping sound that can be quite audible, especially on still evenings and is a useful method of locating beavers in dark conditions or thick vegetation.

Beavers possess a specially adapted dentition for processing their diet (**Figures 18 and 19**). They cut plant material with their incisors and grind it with their molars. Like all rodents their four incisor teeth grow continuously, the hard layer of orange outer enamel (colouration due to iron fortification in the enamel) on their front surface eroding more slowly than the softer, inner layer of dentine, which results in a distinctive chisel-like angle. Beavers will sharpen their teeth by grinding them together and this process ensures that they always possess a sharp cutting edge.

The ability of beavers to digest plant material is aided by fermentation in their hind-gut or caecum. They have a typical rodent digestive system, which consists of a simple stomach with an enlarged

Figure 18 Skull and dentition lateral view.

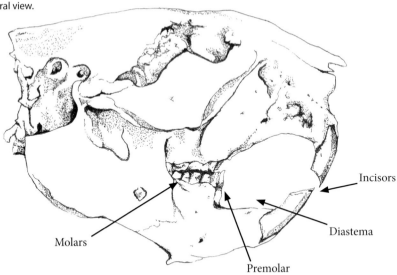

Incisors

Diastema

Molars

Premolar

Figure 19 Skull and dentition ventral view.

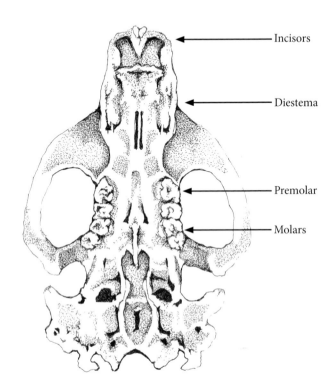

Incisors

Diestema

Premolar

Molars

hind-gut (**Figure 20**). The hind-gut contains a substantial population of micro-organisms (bacteria and protozoa), which enables beavers to ferment cellulose. This process is only 33% efficient in terms of ingested cellulose (Currier *et al.* 1960) and, combined with a relatively short retention time, this means they have to consume large amounts of food. All beavers, except un-weaned kits, display a behaviour called caecotrophy, which results in them licking specially produced green faeces from their anus, which when re-ingested maximises their nutrient uptake.

Beavers tend to avoid coniferous tree species but will sometimes consume smaller saplings or side branches and occasionally fell larger trees for building purposes. In some locations particularly in late winter or early spring they have been recorded ring barking small numbers of conifers growing in close proximity to the waters' edge. While the reasons for this behaviour are not clear it may be that they are seeking the sugars contained in the resin (Krojerová-Prokešová *et al.* 2010).

Beavers are a freshwater species and although they will easily move through salt and brackish water, they must have access to fresh drinking water daily. The ingestion of large amounts of salt, through grooming rather than drinking, can prove detrimental or even fatal in the long term. They drink, usually while swimming, by holding their noses horizontally, submerging their mouths and making chewing motions with their lower jaws (Wilsson 1971).

Figure 20 Digestive system of beavers. Beavers have a simple stomach and digest much of the cellulose ingested in the large intestine and caecum using microbial fermentation (adapted from Vispo and Hume 1995).

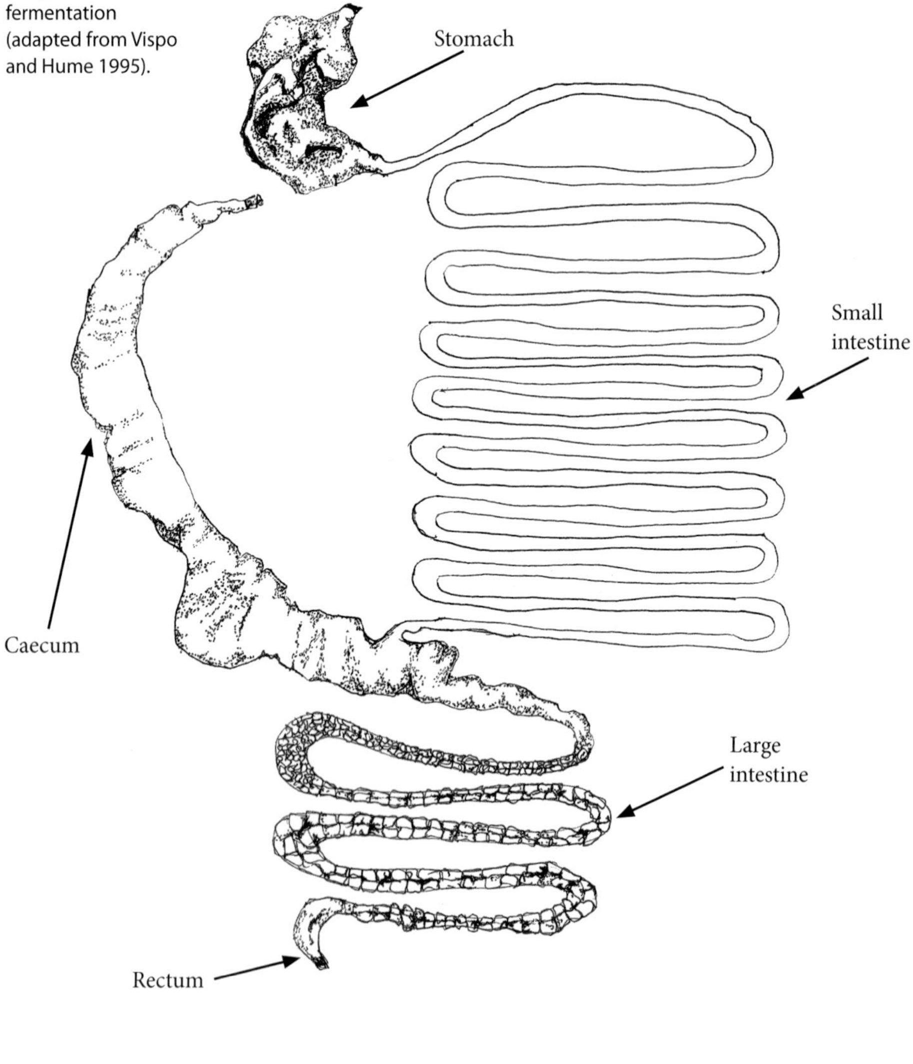

Stomach

Small intestine

Caecum

Large intestine

Rectum

CHAPTER 3

Habitat and Population Biology

Life history

While captive beavers have been recorded as surviving to the age of 28, their average lifespan in the wild is believed to be 12–14 years.

The Eurasian beaver is host to specific parasites (**Table 1**) including the beaver beetle, *Platypsyllus castoris*, a highly modified invertebrate that lives on their fur; the beaver fluke, *Stichorchis subtriquetrus*, an intestinal fluke; and the beaver nematode, *Travassosius rufus*, which is found in their stomachs. These parasites are not transmissible to other species. Both the beaver beetle and the beaver fluke have been recorded in free-living beavers in Scotland (Duff *et al.* 2013; Campbell-Palmer *et al.* 2013).

Although there are no significant natural predators (apart from humans) of adult beavers in Britain, kits are potentially at risk from foxes (*Vulpes vulpes*), badgers (*Meles meles*), dogs (*Canis lupus familiaris*), some eagle species and even large pike (*Esox lucius*). There is growing anecdotal evidence that otters (*Lutra lutra*) may also be opportunistic predators of young kits. Where high population densities exist, dispersing beavers will often be wounded or killed in territorial conflicts with other beavers. The flooding of burrows and lodges in sudden water rises caused by periods of heavy rain for example, is often a cause of high mortality in certain years and areas. Deaths due to road accidents may also become a significant cause of mortality and small numbers of beaver fatalities have been reported on roads in Perthshire, Scotland.

Movements, activity and territoriality

Beaver territories tend to be linear, ranging from 0.5 km (up to 20 km recorded, average 3 km) of shore or riverbank. In very good habitat beaver families can occupy territories of 0.5–0.7 km of bank with ~150–200 m long gaps between adjacent territories (Novak 1987). In poor quality habitat territories can be more widely spaced at 5–10 km intervals. The size and number of territories depends on a number of factors including the density of beaver populations, habitat quality, the number of family members and their settlement pattern (Campbell *et al.* 2005).

Table 1 Diseases of Eurasian beaver (adapted from Goodman et al. 2012).

	Disease	Cause	Transmission	Additional considerations
Bacterial	Tularemia	*Francisella tularensis*	Ingestion of contaminated water	Zoonotic, not present in UK
	Yersina	*Yersina enterocolitica psedotuberculosis*	Contaminated food or water	Zoonotic, present in British wildlife
	Leptosipirosis	*Leptospira* spp.	Contact with infected urine	Zoonotic, present in British wildlife
	Salmonella	*Salmonella* spp.	Ingestion of contaminated food and water	Zoonotic, present in British wildlife
	Campylobacter	*Campylobacter jejuni*	Ingestion of contaminated food and water	Zoonotoc, present in British wildlife
	Toxoplasma	*Toxoplasma gondii*	Contaminated food (shed by domestic cats in faeces)	Uncommon but beaver may act as an intermediate host
Parasitic	Echinococcus	*Echinococcus multilocularis*	Ingestion of eggs from final hosts (foxes/domestic dogs/cats)	Zoonotic, uncommon but beaver may act as an intermediate host. Not currently present in UK
	Giardia	*Giardia lamblia*	Ingestion of contaminated food and water	Zoonotic, present in British wildlife
	Cryptosporidium	*Cryptosporidium parvum*	Ingestion of contaminated food and water	Zoonotic, present in British wildlife
Viral	Rabies	*Lyssavirus* genus	No transmission of rabies from rodents to humans has been documented	Zoonotic, rare, not present in the UK

Beavers are highly territorial with both sexes and all family members participating in defensive behaviours. These principally consist of chemical defence via scent marking but can include physical interactions such as boundary displays to neighbouring individuals, or direct fighting. Perhaps the most interesting of these behaviours in response to the presence of a rival from a different territory is the 'stick display', in which an individual picks up a stick, rises up on to its hind legs and rapidly moves its upper body up and down whilst holding the stick in its mouth and forepaws, in shallow water, creating much splashing (Thomsen *et al.* 2007, **Figure 21**).

Beavers are most active during the hours of darkness but can emerge from their lodges at dusk and dawn in full daylight during the spring and summer months. At this time of year they can be active for 12–14 hours a day. While the time budgets between the sexes do not generally differ for most activities, males tend to exhibit longer daily activity periods as they travel further during territorial patrols (Sharpe and Rosell 2003). Individuals can move up to 5–9 km in a single night (Nolet and Rosell 1994).

From spring onwards the amount of time beavers spend sleeping and resting decreases while grooming and feeding behaviours increase. Breeding adults become warier and more sensitive to disturbance during the birth and emergence seasons for their kits. Lodge maintenance and food caching behaviours are more frequent during the autumn months as beavers prepare for the winter by putting on weight and gathering food caches.

Figure 21 Stick display, a rarely seen intimidation display in beavers, Norway (O. Haarberg).

Beaver offspring remain with their parents and other siblings in their family territories for the first two years of their lives. As they mature they make exploratory excursions into the surrounding environment before finally dispersing to establish a territory and find a partner of their own. This process of dispersal can be delayed by a number of factors such as high population densities, poor habitat quality or a lack of available partners. As a result some offspring may continue to live in a loose, non-breeding association with their original family members for many years.

Habitat

Fresh water habitat types associated with beavers include:

- Large river systems
- Ponds and lakes
- Burns and streams
- Reed beds and marshes
- Areas of deciduous woodland or scrub associated with fresh water
- Ditches with herbaceous plants
- Other riparian features associated with human modified landscapes – including gardens, sewage filter beds/treatment beds, fishing pools, drainage ditches and public parks.

Beavers use water as a refuge from predators, to transport larger food items and to store food over winter (Wilsson 1971). They prefer slow moving water bodies to those with more rapid flows and seldom occupy territories in watercourses with gradients above 2% (Hartman and Tornlov 2006). Beavers require a water depth of at least 0.7 to 1 m to build their lodges and are more likely to dam watercourses which are narrow and/or shallow. Throughout Europe beavers are found in a range of habitat types. They are a highly adaptable species which can capably inhabit intensively managed agricultural landscapes, engineered river systems and even urban environments. In milder climates where suitable vegetation is available all year round they will continue to feed through the winter and therefore do not need to rely on stored caches. Where the climatic conditions are more severe their occupancy of environments is limited by the availability of woody material to establish winter food caches.

CHAPTER 4

Beaver Field Signs

Beavers can be quite secretive and direct observations of them can be limited by their semi-aquatic lifestyle and nocturnal behaviours. In large open water bodies they can be relatively easy to spot at dawn and dusk while in small, narrow stream systems they can be very difficult to locate and observe. Luckily they leave very obvious field signs, which are usually the first indication of their presence (see **Appendix 2**). Recording these field signs can help to identify their wider distribution and allow an assessment of their habitat use. This information can then be expanded to afford an estimation of the number of active territories present within an area.

Teeth marks

Beavers leave distinctive markings on the woody vegetation they feed upon or fell (**Figures 22 and 23**).

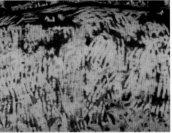

Figure 22 Clear teeth marks left by beavers (Scottish Beaver Trial).

Figure 23 Clear teeth marks left by beavers (R. Campbell-Palmer).

Felled and gnawed trees

The tooth pattern left in beaver felled wood is highly distinctive and is sometimes still visible in remains from archaeological sites (Coles 2006). On smaller side branches the angle of their cut results in a 'flute' shaped profile, while on larger tree trunks their gnawing and felling can result in very distinctive 'pencil sharpened' points. Marks left by their incisors can be clearly seen or felt by running a finger across the cut end (**Figures 22 and 26**). When beavers gnaw through substantial timber they produce distinctive chippings (**Figures 24 and 25**). In common with their other feeding signs these wood chips can be aged by their colouration, which are light when fresh and darken with age (**Figure 26**). Beavers are untidy feeders and many of their freshly peeled sticks float away on the surface of the water into the wider environment. In habitats which are densely vegetated or

Figure 24 Partially felled tree (left, R. J. Needham).

Figure 25 Distinctive beaver chippings (top right, R. J. Needham).

Figure 26 Felled tree stump over a year old: ridges created by teeth can still be felt by running a finger over the cut end (bottom right, P. Creech).

extensive, identification of these pale twigs in the water or on land along the shoreline can often be the first indication of the presence of beavers.

Grazed lawns /vascular plants

Beavers will commonly feed on a nightly basis on areas of terrestrial vegetation adjacent to a watercourse. In time this activity can produce closely cropped lawns (**Figures 27 and 28**) with characteristic

Figure 27 Grassy area with beaver grazed section in centre of image (Derek Gow Consultancy Ltd).

Figure 28 Typical beaver grazed crops (Derek Gow Consultancy Ltd).

Figure 29 Typical beaver cut stem (left, Derek Gow Consultancy Ltd).

Figure 30 Beaver feeding stations can range greatly in size and shape but sticks with bark peeled from them and beaver cut ends will be evident (right, Scottish Beaver Trial).

features which can be distinguished from similar lawns created by waterfowl through their absence of droppings and feathers. Beaver feeding on the stems of larger, rigid plant species can result in a distinctive ~45° cut (Figure 29). On smaller, lighter vegetation such as soft rush (*Juncus effusus*) this cut pattern can be very similar to the feeding signs of water voles (*Arvicola amphibious*).

Feeding stations

Beavers often repeatedly feed at the same location on the water's edge for extended periods before moving on to another area. The vegetation debris that accumulates is often visible at the water's edge. Piles of peeled, pale, discarded sticks can provide a good indication of both beaver presence and fresh activity (Figure 30). These feeding stations can vary greatly in size from just a few sticks to well over 100. Their approximate age can be established by the colouration of the peeled sticks (lighter tends to equal fresher) and/or the presence of freshly cut green vegetation.

Figure 31 Foraging trail demonstrating dam and canal construction (R. J. Needham).

Foraging trail between two water bodies

Dam construction creating canal

Foraging trails

Where beavers repeatedly forage on land in a specific area their regular movements create trails and pathways (**Figure 31**). These features are further worn by the repetitive dragging of branches back to the water. Following these trails inland may lead to further field signs such as cut branches and felled or gnawed trees. Although otters, deer and other mammals will also create well-worn trails, the semi-excavated character of beaver trails is highly distinctive. These features make excellent locations for positioning camera traps.

Lodges and burrows

Beaver burrows or lodges are the focal points of a beaver family. They can be created by pairs or sometimes even single individuals. Lodges provide shelter from the elements and predators. They can

Figure 32 Typical development of a bankside lodge, beginning with a burrow into the bank (adapted from Coles 2006).

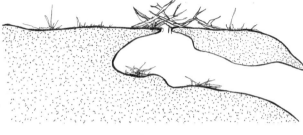

Figure 33 Further burrows and chambers can then be dug out (adapted from Coles 2006).

Figure 34 Additional collected material may be piled on top, within which chambers may be gnawed out. Note water levels of associated water body may fluctuate, but burrow entrance tends to be continually submerged, with upper burrow and chambers being above the water line (adapted from Coles 2006).

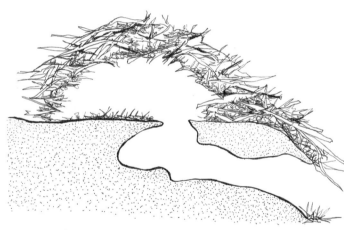

vary greatly in shape and size depending on the surrounding habitat and number of occupants. Some dwellings can be very obvious structures at the water's edge, while others mainly consist of burrows or chambers with few obvious external features. For example, where flooding is common lodges can extend for many metres into a bank. These structures can contain a number of chambers at different levels which will be alternatively utilised by beavers depending on the external water levels. Construction materials tend to consist of branches, mud and other vegetation (Willson 1971).

To create a lodge (**Figures 32, 33 and 34**) beavers first tend to dig a burrow or excavate a cleft in banking next to the water. In rocky environments or artificially reinforced river banks beavers may adapt any available shallow shelves. Once a construction point is selected it is built upon or added to by piling layers of sticks, mud and vegetation on top. Most lodges have two entrances, which are below the water level and seldom visible. Where branches protrude into the walls of the chamber they tend to be smoothed off through gnawing while the internal floor is lined with a bed of finely split wood shreds. Their living chambers tend to be formed in one of two ways. If the bankside substrates are friable then the beavers will excavate tunnels and chambers by digging. If this is not possible they will build outwards and upwards, creating a 'roof' using woody vegetation and mud before creating chambers and tunnels inside this by gnawing. Lodge maintenance behaviours are most prevalent in the autumn as beavers prepare for the winter. Signs of recent activity include wet mud or aquatic plants deposited on the outer surface. Branches with green leaves which have not yet wilted are good indicators of fresh activity.

There are two principal lodge designs; a freestanding structure (**Figure 35**) in a pool, or a bankside lodge. The former type is established either by building directly on top of the shoreline or when beavers dam a small stream to create a pool which then expands to surround the lodge. This is then expanded in response to the deepening water. A bankside lodge (**Figure 36**) is a similar structure, but is located on

Figure 35 Free-standing lodge in autumn (R. J. Needham).

Figure 36 Bankside lodge with winter food cache in the water (front right) (Scottish Beaver Trust).

the water's edge and formed around a burrow dug into the bank or created from material extending from the shoreline.

Beavers do not necessarily live in lodges all year round. They are excellent burrowers and can quickly create 'day rests' which are simple 'short' burrows or day beds, and they may sleep here instead of the main lodge in warmer months. In environments with friable soil banks they prefer to burrow rather than build lodges. Their living chambers are always above the normal height of the water with submerged access tunnels. Although some of these structures can extend back into a riverbank for over 20 m most are <5 m in length. Sometimes burrows also act as access routes from the water's

Figure 37 Beaver day bed (J. Coats).

edge to food sources further up the bank. At times upper sections of these tunnels may partially collapse, exposing the burrows and chambers beneath. If the beavers are still in residence when this occurs they will commonly reinforce these areas with sticks and mud. Burrows are commonly invisible above ground although their entrances can sometimes be observed where water visibility is clear or where levels are unusually low. In narrow watercourses these burrows are most likely to be found by viewing from the opposite bank or from a canoe in wider or deeper watercourses. Undertaking searches in the autumn and winter is recommended as field signs are easier to locate once the growing vegetation has died back. The growth of tall vegetation in the summer months can easily obscure even large beaver lodges.

Dams

Beavers will not build dams in rivers and lakes, but they will readily dam small streams and ditch systems. Beavers dam to retain and

Figure 38 Multiple impoundments and pools (R. Campbell-Palmer).

Figure 39 Early
stages of dam
construction (R. J.
Needham).

manage water levels. The pools which result from damming ensure that the entrances to their living quarters remain submerged allowing them to enter and exit safely. The deeper water retained by a dam also affords submerged access to food caches should the surface become frozen in winter. Beavers prefer to move through water as movement overland makes them more vulnerable to predators.

Dams are constructed from various materials including sticks, branches, mud, stones and vegetation (**Figure 39**). Beavers skilfully integrate these into tightly interwoven structures (**Figures 38 and 40**) and will utilise natural features such as channel narrowing, fallen trees or large rocks to assist this process. When complete, beaver dams are fronted on their upstream side by ramps of captured silt or specifically positioned debris. This design ensures that they are highly resilient to flood conditions. Non-beaver related 'debris' dams occur in streams where detritus is trapped against an obstacle in the watercourse. These dams can be distinguished from beaver dams by a lack of uniformity in their structure and an absence of beaver cut sticks.

It should be noted that the creation of dams by beavers depends on habitat type. For example, if they occupy lakes, gravel pits or wide rivers (>10m) with friable banks and an abundance of easily accessible food they tend not to build dams. By comparison, beavers living on narrower water bodies can create extensive systems of multiple dams and impoundments.

Canals

Although well-worn foraging trails may eventually fill with water, beavers will also actively excavate networks of small canals (**Figures 41 and 42**) in flatter habitats. These structures, which radiate out from a principal water body, will be cleared by the beavers of sediment, detritus, fallen leaves or other vegetation on a regular basis. This excavated debris is placed on the canal side in irregular mounds. Beavers use these canal features to transport browse and building materials and they also provide escape routes back to deeper water and safety.

Where conditions allow, numerous canals can be created in a beaver territory. Most are of short length but some can extend 150 m or more

Figure 41 Beaver
created canal, leading
from main water
body (R. J. Needham).

from the main water body, with an average width of 40–50 cm. These canals open up new feeding areas for the beavers and create access corridors that are also exploited as habitats by many other species. In times of drought beavers will create small dams or muddy impoundments in canals to maintain a basic water level and these features afford an important resource for a range of other wildlife.

Figure 42 Beaver eye view of canal (~40 cm in width) (R. J. Needham).

Figure 42 Beaver eye view of canal (~40 cm in width) (R. J. Needham).

Scent mounds

Figure 43 Beaver scent mound (R. Campbell-Palmer).

Scent mounds (**Figure 43**) are small piles of mud, vegetation or other detritus that beavers gather from the surrounding environment. These are deposited on the bankside or promontory of a water body, and are marked using castoreum and/or AGS. Scent mounds are used by family units to identify their territory and to ward off potential challengers. April and May are the most active months for scent mound creation as dispersing individuals search for new territories. Although scent mounds can be hard to spot, the camphor-like smell associated with these structures is quite distinctive and once recognised easy to remember!

Faeces

Normally beaver urination and defecation occurs in the water (Wilsson 1971). This behaviour makes faeces surveys an unreliable monitoring method. The fibrous nature of their faecal content ensures that breakdown is relatively swift especially in habitats with flowing water. Sunken faeces may occasionally be seen lying on the bottom of the water body in the vicinity of lodges or feeding stations in areas of still or slow moving water (**Figure 44**). Occasionally faecal pellets are also washed up on banks.

Figure 44 Beaver faeces floating in shallow water (Derek Gow Consultancy Ltd).

Tracks/prints Despite beaver tracks being unique field signs they can be hard to find. Muddy or snowy trails, banks with sparse vegetation and areas around their lodges, dams or canals are all good places to search for beaver tracks (**Figures 45 and 46**). Beaver tracks are often distorted by their tails or branches dragging behind them, so intact prints are not always observed. Intact prints of the fore and hind paws differ greatly and can be a great identifying feature. The hind paws are large and long, a clear print will display the outline of the web of skin connecting the toes. It is unlikely that this could be mistaken for any other native riparian mammal in Britain. The smaller forepaw print is also distinctive although the outline of every toe is not always clear, with three or four digits being commonly most visible. Where beaver trails are well used, individual tracks become less distinct.

Figure 45 Beaver tracks in soft mud, forepaw (left) (R. Campbell-Palmer)

Figure 46 Beaver tracks in soft mud, hind paw (right) (R. Campbell-Palmer).

CHAPTER 5

Observing Beavers

Regular observations of beavers at active lodges can occur without disturbance, and without the use of artificial lights, providing the observer remains quiet and still. Lodges located on larger bodies of water such as rivers or lakes are easier to observe from either the opposite bank or from the same bank further up or down the shore. Sitting on top of or very close to the lodge is likely to deter the beavers from leaving or returning and therefore should be avoided. Beavers will often emerge before sunset during the spring and summer months and can be highly visible as they patrol their territorial boundaries or feed along the shoreline. In the autumn they can provide a great viewing spectacle whilst they transport woody vegetation from their foraging areas to create their winter food caches. Beaver watching times can vary depending on their location in Britain. During spring and summer in Scotland an ideal time would be ~18:30–21:00 until nightfall and ~05:00–07:00. In southern England beavers will appear in the late afternoon from ~16.30 onwards and will return to their lodges between ~06:00–07:00. Like watching any other wild animal, patience and local knowledge are essential for increasing the chances of success.

When watching a lodge it is common for beavers to emerge in the water several metres away from the front of the lodge. This is often silent with no prior warning other than a few bubbles. In still water the appearance of significant waves, especially near the shoreline or from under overhanging vegetation, often indicates the presence of a beaver. Beavers will commonly float in the water for a couple of minutes assessing their surroundings before swimming off purposefully to patrol their territory or reach a favoured feeding spot. Rainy or very windy weather conditions are likely to reduce successful observations as beavers are less visible in rougher waters and behave more cautiously. Beavers can often be observed reaching up to pull down and feed on branches over-hanging the water, feeding on emergent vegetation lining the shore or repeatedly diving to forage on submerged aquatic plants. Identifying fresh activity such as a well-used trail, canal or feeding spot may provide potential viewing sites to stake out. Observations from July onwards may be rewarded

Figure 47 Adult (>3 years) swimming: only the head is visible above the water surface (R. J. Needham).

Figure 48 Young beaver swimming: animal appears buoyant with back clearly visible (S. Gardner).

by the sight of newly emerged kits. During their first few weeks of emergence kits will often spend time in the immediate vicinity of the main lodge under the watchful presence of other family members who may usher them back into the lodge should a potential threat become apparent. The kits generally remain within a safe retreating distance while they learn to forage for themselves and perfect their diving behaviours. After a few weeks they become much more independent and will readily head off on their own.

Identification of individual beavers can be difficult but differences in size, particularly in visible body length and width of head, may afford a more detailed assessment. Adults have broad heads and whilst swimming their backs are generally not visible above the water line (**Figure 47**) unless they are floating. Kits (< one year old) swim higher in the water and are quite buoyant. Their backs are completely visible above the water line (**Figure 48**) and their fur may appear almost dry.

Although familiarity gained from regular observations will increase the ability to recognise these individual differences, they are only slight and beavers over the age of ~three years are usually very similar in size. One benefit of watching an active lodge at emergence times is that it may be possible to count the occupants as they come and go (be careful not to double count returning animals!). The reproductive status of the colony can also be determined as if more than two animals are regularly seen in an area bringing food back to the lodge this may indicate the presence of offspring and therefore active breeding.

Remote camera trapping

Camera traps are a non-invasive monitoring method that has many benefits and can be an important source of information for those wishing to study beaver activity and behaviours. They can often be the most effective tool to date for recording more secretive behaviours such as scent marking, canal excavation or lodge and dam maintenance. For such monitoring, cameras should be placed in areas of fresh activity such as dams, lodges, feeding stations (**Figure 49**). Routes where beavers travel on a regular basis such as foraging trails and canals are also good sites to select. Camera trapping can provide information on activity patterns (recorded from the date and time the image was captured), different behaviours and physical characteristics that may enable identification of individual beavers (**Figures 50 and 51**). Permission from the landowner should always be obtained before cameras are placed. Their location should be

Figure 49 Camera trap placed at the beginning of a foraging trail overlooking the land and the water. The arrow shows the angle of the camera lens. Note the large feeding station along the water's edge (R. J. Needham).

carefully considered to ensure that images of people are avoided and to reduce the chances of theft or damage.

Camera traps with a video function are preferable to cameras that only permit still images. The video function enables the recording of the animals' activity over a short period of time. Video also allows the beaver to move within the frames increasing the chance of getting a view of their tail at an angle that may permit the recognition of individuals from unique features such as scars and notches.

Figure 50 Adult beaver displaying interest in a set beaver trap; time of activity and body condition are useful information here (Scottish Beaver Trial).

8/07/2013 8:11 PM

Figure 51 Unique tail 'notching' and scarring can enable individual identification (TUC and Scottish Beaver Trial).

CHAPTER 6

Beavers in Modern Landscapes

Beavers and the law

The current legal status of beavers in Britain is not clear cut, but should wider reintroductions occur in the future then this situation will need to be resolved by the statutory authorities. Releasing beavers into the wild without a government licence is currently an offence under section 16(4) of the Wildlife and Countryside Act (1981) as the species is not considered to be 'ordinarily resident' in Britain. The definition of what constitutes 'ordinarily resident' is unclear at present. For example the current free-living, unlicensed Tayside beaver population in Scotland appears to have been producing wild born offspring for a number of generations. Natural England has suggested that one interpretation of 'ordinarily resident in Great Britain in a wild state' could mean where a release into the wild has resulted in an established population of a species which could be considered to be 'ordinarily resident', then 'subsequent releases would not require a licence'. In Scotland the statutory bodies currently consider beavers not to be resident. The Eurasian beaver is currently listed as a European Protected Species on Annex IV of the EC Habitats Directive, and is also protected under the Bern Convention as an Annex III species. However, beavers are not currently protected by UK legislation as they are not on Schedule 2 of the Habitats Regulations and would remain so unless UK governments take the decision to allow further licensed reintroductions (Gurnell *et al.* 2009). This interpretation has not been legally challenged in Britain, nor has the lethal control of beavers, but neither have there been, to date, any prosecutions over suspected, unlicensed beaver releases into the wild. While the Scottish Government licensed the Scottish Beaver Trial in 2008 and agreed in 2012 to tolerate the Tayside population until a decision on the future of beavers in Scotland is made, they have also been clear that at the time of writing beavers can be culled by landowners without danger of prosecution (Jones *et al.* 2013).

Beavers as ecosystem engineers

Beavers are renowned for their ability to modify the habitat in which they live. An extensive body of scientific research indicates that beaver generated landscapes contain higher levels of biodiversity and biomass on a landscape scale than environments from which

Figure 52 Beaver
created landscape,
Grub Bavaria (Derek
Gow Consulatancy
Ltd).

Figure 52 Beaver created landscape, Grub Bavaria (Derek Gow Consulatancy Ltd).

they are absent (Rosell *et al.* 2005), However there can be 'winners' and 'losers' in other groups of species generally when beavers colonise new habitats. In British riparian landscapes, which have historically seen widespread and prolonged declines due to intensive drainage and agricultural practises, the conservation value of beaver activities in restoring lost wetlands could be significant in the long term. Should we decide to live alongside beavers once more in the future then our modern landscapes and perceptions of traditional woodland and freshwater management will be challenged, away from uniform and tidy forests and canalised waterways. Beaver activities on a landscape scale trigger the development of ecological cascades. This dynamic process restores areas of standing and submerged deadwood habitats to riverine woodlands and reinstates irregular water courses with an abundance of associated pools, wetlands and swamp forests (**Figure 52**).

The return of the beaver

By the late nineteenth century the Eurasian beaver had been hunted to the point of almost extinction for its fur, meat and castoreum. At their lowest point it is believed that the Eurasian beaver was reduced to ~1,200 individuals scattered in isolated populations across Eurasia. Some of these populations included ~200 individuals on the River Elbe in Germany, ~30 on the River Rhone in France and ~100 in the Telemark region of Norway (Nolet and Rosell 1998). Since the

1900s, beaver numbers have recovered throughout much of their former range through protective regimes, hunting regulation, active reintroductions and natural recolonisation. The first beaver translocation, from Norway to Sweden, occurred in 1922, and since then, there have been more than 205 recorded translocations which have returned beavers to 25 nations where they were formerly extinct. There are currently estimated to be approximately 1.04 million Eurasian beavers distributed throughout much of their former native range (Halley *et al.* 2012) (**Figures 53 and 54**).

Figure 53 European distribution of Eurasian beavers marked in red, historic populations marked in black and North American beaver populations marked in green (Halley *et al.* 2012).

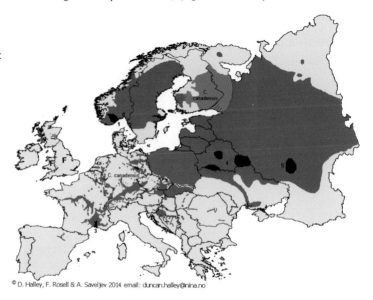

© D. Halley, F. Rosell & A. Saveljev 2014 email: duncan.halley@nina.no

Beaver releases have been undertaken for a wide variety of reasons, ranging from a desire to re-establish a harvestable fur resource to the increasing desire to see beavers restored to their natural function as effective managers of wetland landscapes. Beavers create landscapes which retain more water, which may play an important role in regulating floods, retaining water during droughts and reducing the impacts of flooding on human communities (Jones *et al.* 2013). These 'ecosystem services', largely through their dam building behaviours, could help to alleviate flood events by the slowing of peak discharges and increasing water retention. Beaver dams can also trap sediments, carbon, toxins and excess nutrients draining from surrounding intensively managed landscapes (Hood and Bayley 2008; Pope 2013).

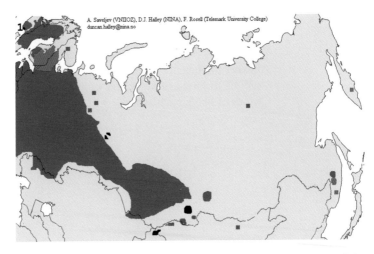

Figure 54 Asian distribution of Eurasian beavers marked in red, historic populations marked in black and North American beaver populations marked in green (Halley *et al.* 2012).

Re-establishing an appreciation of the ecological importance of the beaver through educational initiatives will be critically important to ensuring its future coexistence with people. Beavers are living wild in a number of locations throughout Britain with some of these animals being escapees (or their progeny) of captive collections (Jones *et al.* 2013). It is also possible that some small-scale illegal releases may have resulted from individuals who wish to see the species restored. Private landowners, various stakeholders, government agencies, the general public and non-governmental organisations are all therefore interested in the presence of beavers and their activities.

Over the past 40 years there has been a steady growth in the number of both officially approved and unsanctioned beaver restorations in Europe. In Belgium and Spain illicit releases of beavers have established free-living populations, which are now expanding. Modern European experience clearly demonstrates that beavers can adapt to contemporary cultural landscapes. In certain areas their existence is compromised by habitat loss or degradation and the spread of introduced North American beavers is of real concern (Parker *et al.* 2012). The main threat currently facing the Eurasian beaver throughout most of its expanding range is the increasing potential for their activities to conflict with human land-use. In many countries beavers have been absent or confined to tiny relict populations for many centuries. Developing a re-understanding of their abilities, needs and requirements can often be culturally difficult for some land-users and the organisations that represent them. Proponents of beaver reintroduction in Britain must conversely understand that their future existence will only prove practicable within tolerable limits. Over time forms of beaver management including both lethal and non-lethal control will be required to develop a wider acceptance of this species in a modern British landscape.

Acknowledgements

We would like to thank all those who contributed to the information, diagrams and photographs used in this booklet, including: Andy Beer, Rachael Campbell-Palmer, Bryony Coles, Steve Gardner, Simon Girling, Gidona Goodman, Duncan Halley, Andrew Kitchener, Jo McIntrye, Howard Parker, David Plummer, Romain Pizzi, Phillip Price and Gerhard Schwab. Special thanks to John Gurnell for his constructive comments on earlier drafts.

This booklet would not have been possible without sponsorship from the following organisations:

THE MAMMAL SOCIETY

THE ROYAL ZOOLOGICAL SOCIETY OF SCOTLAND

OWNERS OF EDINBURGH ZOO & HIGHLAND WILDLIFE PARK
REGISTERED CHARITY No. SC004064

Scottish Wildlife Trust

SCOTTISH BEAVER TRIAL

Telemark University College

Appendix 1
Wild plants commonly eaten by beavers.

Common name	Latin name
Alder	*Alnus glutinosa*
Angelica	*Angelica sylvestris*
Ash	*Fraxinus* spp.
Aspen	*Populus* spp.
Aster, sunflower	Asteraceae family
Bedstraw	*Galium* spp.
Beech	*Fagus* spp.
Bellflower	*Campanula* spp.
Birch	*Betula* spp.
Blackberry	*Rubus fruticosus*
Bog Bean	*Menyanthes trifoliata*
Buckbean	*Menyanthes trifoliata*
Butterbur	*Petasites* spp.
Bracken	*Pteridium aquilinum*
Branched Bur-reed	*Sparganium erectum*
Canary grass	*Digraphis arundiancea*
Cattails	*Typha latifolia*
Cherry, plum, peach, almond	*Prunus* spp.
Cinquefoils	*Potentilla* spp.
Clover	*Trifolium* spp.
Common club-rush	*Schoenoplectus lactustris*
Crowfoot family	Ranunculaceae
Dandelion	*Taraxacum* spp.
Downy burdock	*Arctium tomentosum*
Dropwort	*Filipendula vulgaris*
Elm	*Ulmus* spp.
Geum	*Geum rivale*
Goldenrod	*Solidago virgaurea*
Ground elder	*Aegopodium podagraria*
Hawthorn	*Crataegus*
Hazel	*Corylus* spp.
Himalayan Balsam	*Impatiens glandulifera*
Horse sorrel	*Rumex*
Iris	*Iris pseudacorus*
Japanese knotweed	*Fallopia japonica*
Maple/horse chestnut	*Acer* spp.
Mint family	Lamiaceae

Common name	Latin name
Legume family	Fabaceae
Lime	*Tila cordata*
Lizard's tail	*Saururus cernuus*
Marsh cinquefoil	*Potentilla palustris*
Marsh marigolds	*Caltha palustris*
Meadowsweet	*Filipendula ulmaria*
Mugwort	*Artemisia vulgaris*
Nettles	*Urtica* spp.
Oak	*Quercus* spp.
Pondweed	*Potamogeton*
Poplar	*Populus* spp.
Raspberry	*Rubus idaeus*
Red dogwood	*Swida sanguinea*
Reeds	*Phragmites*
Reed grass	*Calamagrostis lanceolatus*
Rowan	*Sorbus aucuparia*
Rushes	*Juncus* spp.
Saw sedge	*Cladium mariscus*
Sedges	*Carex* spp.
Silverberry	*Elaeagnus commatata*
Sweet flag	*Acorus calamus*
Sycamore	*Acer pseudoplatanus*
Thistles	*Cirsium* spp.
Timothy	*Phleum pratense*
Velvet plants	*Verbascum*
Water avens	*Geum rivale*
Water horsetail	*Equisetum fluviatile*
Water lilies	*Nymphaea alba* and *Nuphar lutea*
Water milfoil	*Myriophyllum* spp.
Waterweeds	*Elodea* spp.
Willow	*Salix* spp.
Willow herb	*Epilobium angustifolium*
Witch hazel	*Hammamelis virginiana*
Yarrow	*Achillea millefolium*

Adapted from Kitchener (2001), additional information from Willby *et al.* (2011) and observations of free living Scottish beavers.

Appendix 2

Beaver field signs, features and possible confusions.

Field signs	Identification	Possible confusion
Teeth marks	• Distinct teeth marks • Ridges can be felt for by running a finger over the cut end	
Felled and gnawed trees	• Felling cut at ~45° • Distinct teeth marks • Wood chipping around base of tree	• At a distance trees fallen and snapped in wind can be mistaken for felled trees • Bark stripping by deer may be mistaken for gnawing by beavers
Peeled woody material	• Bark peeled from sticks • Teeth marks very often visible • Stick ends very distinct	
Angled cuts on vascular stems	• Clean cut at ~45° • No frayed edges	• Lagomorphs • Water vole
Grazed lawns/crops	• Areas of lawns/crops grazed with clear foraging trails leading to and from the water	• Lagomorphs
Feeding stations	• Piles of peeled sticks • Varying quantity of sticks (2–100+) • Varying sizes of sticks • Age assessed by the colouration of sticks (very fresh is light in colour gradually getting darker with age)	• Natural stick piles/debris
Forage trails	• Flattened vegetation pathways • Always leading from the water • Any other beaver feeding signs along the trail	• Other mammals such as otters and deer
Lodges and burrows	• Large piled structure made from sticks, mud and vegetation • Varies greatly in size and shape • Fresh construction activity prevalent in autumn and spring • Fresh signs include wet/damp mud or vegetation	• Burrows can be very hard to find as the entrance will be under the water. Can be found in low water conditions

Field signs	Identification	Possible confusion
Dams	• Linear structure across a water course • Interwoven from sticks, mud and vegetation	• Debris dams can be distinguished by the lack of uniformity and beaver cut material
Canals	• Narrow channel excavated; often originate from forage trials • Radiate from the primary water body • Excavated debris deposited on canal sides	• Artificial drainage ditches in Forestry and Agricultural land
Scent mounds	• Small piles of mud or detritus, near water's edge and territorial boundaries, marked with distinctive scent	
Faeces (scats)	• Not often seen as beavers defaecate in water, might be seen in slow moving water around lodges, dams, feeding stations	• Otter spraints, but can be distinguished through smell and presence of food remains
Tracks	• Hind paws are large and long • Possible to see the print of the web of skin connecting the toes on hind feet • Distorted tracks possible due to tail being dragged through the tracks	• Distorted tracks may be confused with other riparian mammals

Appendix 3
Ethogram of common beaver behaviours.

Behaviour	Specific	Description
Locomotion	Swim	Most swimming occurs on water surface (head visible). Swimming underwater mainly when disturbed, entering lodge or foraging on aquatic plants
	Walk	On all fours, very short distances bipedally when carrying objects in forelegs
	Dive	Often silent; base of back and tail may be seen at water's surface before disappearing underwater
Feeding	Forage	Gathering of vegetation on land, water's edge or underwater; will often carry material in mouth to take to a preferred location to feed on
	Eat	Often quite noisy chewing; will hold food item in forepaws
	Forage: aquatic	Often seen as a series of short dives in a small area, will carry vegetation to surface to take into shore area to feed or can remain in the water holding food in forepaws and feed with head and mouth just above the waterline
	Food cache	Tends to involve the cutting and carrying of non-peeled woody material for storage in water outside lodge
	Bark stripping	Removal of bark from branches/sticks/tree trunks
Territorial/ defensive	Scent-mark	Scent tends to be deposited on specifically made mounds of sediment/vegetation, often at territory borders
	Fight	Tends to occur with unrelated beavers, often involving biting around the rump and tail region
	Tail slap	Whilst in water, tail is raised out of water and brought down fast onto water surface to make a loud and distinctive noise
	Patrol	Monitoring territory borders, sniffing and scent marking commonly included

Behaviour	Specific	Description
Social	Play	Tends to be seen in younger animals; can involve short bouts of wrestling like behaviours, climbing on top of each other, stealing items from each other etc
	Wrestle	Often seen in water between two individuals, head on, pushing with forearms
	Allogroom	Grooming of family member with teeth; forepaws may be rested on the other animal for contact and balance. Reinforces family bonds
	Vocalisation	Younger animals tend to be most vocal; older animals may vocalise when they meet when out and about. Threatened beavers may hiss and grind teeth
Habitat modification	Build	Intricate movement and construction of felled material/sediment/vegetation into specific structures such as dams and lodges
	Dig	Construction of burrows/canals with forepaws
	Felling	Gnawing of trees, tending to hold head at an angle, raised on back legs with forepaws resting on trunk
	Channel deepening	Digging and gnawing at material on bottom of a beaver canal/forage trail and the piling of lose debris on either side of the resulting channel

Produced with reference to Wilsson 1971. Beaver observations by Scottish Beaver Trial and TUC staff.

Appendix 4

Useful information sources:

Natural England
1 East Parade, Sheffield S1 2ET
Tel: 0300 060 6000
www.naturalengland.org.uk

The Mammal Society
3 The Carronades, New Road, Southampton SO14 0AA
Tel: 023 8023b 7874
www.mammal.org.uk

Royal Zoological Society of Scotland
134 Corstorphine Road, Edinburgh, EH12 6ST
Tel: 0131 334 9171
www.rzss.org.uk
Email rcampbellpalmer@rzss.org.uk

Scottish Beaver Trial
Cairnbaan Forestry Office, Cairnbaan, Lochgilphead, PA31
Tel: 01546 603346
www.scottishbeavertrial.org.uk

Scottish Natural Heritage
Great Glen House, Leachkin Road, Inverness IV3 8NW
Tel: 01463 725000
www.snh.gov.uk

Scottish Wildlife Trust
Harbourside House, 110 Commercial Street, Edinburgh EH12 6NS
www.scottishwildlifetrust.org.uk

Telemark University College
Department of Environmental and Health Studies, Telemark
University College, N-3800 Bø i Telemark Norway
Email Frank.Rosell@hit.no
www.hit.no/ansatte/vis/frank.rosell

Tayside Beaver Study Group
www.taysidebeaverstudygroup.org.uk

References

Cambrensis, G. (1188) *Itinerarium Cambridae (A Journey Through Wales).*

Campbell, R. D., Rosell, F., Nolet, B. A., Dijkstra, V. A. A. (2005) Territory and group size in Eurasian beavers (*Castor fiber*): Echoes of settlement and reproduction. *Behaviour Ecology and Sociobiology* **58**: 597–607.

Campbell-Palmer, R. and Rosell, F. (2010) Conservation of the Eurasian beaver *Castor fiber*: An olfactory perspective. *Mammal Review* **40**: 293–312.

Campbell-Palmer, R., Girling, S., Pizzi, R., Hamnes, I. S., Øines, Ø. and Del-Pozo, J. (2013) *Stichorchis subtriquestrus* in a free-living beaver in Scotland. *Veterinary Record* **173**: 730.

Coles, B. (2006) *Beavers in Britain's Past.* Oxbow Books, UK.

Currier, A., Kitts, W. D. and Cowan, I. (1960) Cellulose digestion in the beaver (*Castor canadensis*). *Canadian Journal of Zoology* **38**: 1109–1116.

Duff, A. G., Campbell-Palmer, R. and Needham, R. (2013) The beaver beetle *Paltypsyllus castoris Ritsema* (Leiodidae: Platypsyllinae) apparently established on reintroduced beavers in Scotland, new to Britain. *The Coleopterist* **22**: 9–19.

Goodman, G., Girling, S., Pizzi, R., Meredith, A., Rosell, F. and Campbell-Palmer, R. (2012) Establishment of a health surveillance program for reintroduction of the Eurasian beaver (*Castor fiber*) into Scotland. *Journal of Wildlife Diseases*, **48**: 971–978.

Gurnell, J., Gurnell, A. M., Demeritt, D., Lurz, P. W. W., Shirley, M. D. F., Rushton, S. P., Faulkes, C. G., Nobert, S. and Hare, E. (2009) *The Feasibility and Acceptability of Restoring the European Beaver to England.* Natural England Commissioned Report NECR002.

Haarberg, O. and Rosell, F. (2006) Selective foraging on woody plant species by the Eurasian beaver (*Castor fiber*) in Telemark, Norway. *Journal of Zoology* **270**: 201–208.

Halley, D., Rosell, F. and Saveljev, A. (2012) Population and distribution of Eurasian beavers (*Castor fiber*). *Baltic Forestry* **18**: 168–175.

Hartman, D. J. and Tornlov, S. (2006) Influence of watercourse depth and width on dam building behaviour by Eurasian beaver, *Castor fiber. Journal of Zoology* **268**: 127–131.

Hood, G. A. and Bayley, S. (2008) Beaver (*Castor canadensis*) mitigate the effects of climate on the area of open water in boreal wetlands in western Canada. *Biological Conservation* **141**: 556–567.

Horn, S., Durka, W., Wolf, R., Ermala, A., Stubbe, A., Stubbe, M. and Hofreiter, M. (2011) Mitochondrial genomes reveal slow rates of molecular evolution and the timing of speciation in beavers (*Castor*), one of the largest rodent species. *PLOS One* **6**: e14622.

Irving, L. (1937) The respiration of beaver. *Journal of Cellular and Comparative Physiology* **9**: 437–451.

Jones, S., Gow, D., Lloyd Jones, A. and Campbell-Palmer, R. (2013) The battle for British Beavers. *British Wildlife* **24**: 381–392.

Karraker, N. E. and Gibbs, J. P. (2009) Amphibian production in forested landscapes in relation to wetland hydroperiod: A case study of vernal pools and beaver ponds. *Biological Conservation* **142**: 2293–2302.

Kitchener, A. (2001) *Beavers*. Whittet Books, UK.

Krojerová-Prokešová, J., Barančeková, M., Hamšiková, L. and Vorel, A. (2010) Feeding habits of reintroduced Eurasian beaver: Spatial and seasonal variation in the use of food reserves. *Journal of Zoology* **281**: 183–193.

Lamsodis, R. and Ulevicius, A. (2012) *Impact of Beaver Ponds on Migration of Nitrogen and Phosphorous via Drainage Ditches in Agrolandscapes, Lithuania*. Poster 6th International Beaver Symposium, Ivanic-Grad, Croatia.

Lavrov, L. S. and Orlov, V. N. (1973) Karyotypes and taxonomy of modern beavers (*Castor*, Castoridae, Mammalia). *Zoologicheskii-Zhurnal* **52**: 734–742.

Martin, L. D. (1994) *The Devil's Corkscrew*. http://www.naturalhistorymag.com/htmlsite/master.html?http://www.naturalhistorymag.com/htmlsite/editors_pick/1994_04_pick.html

McKean, T. (1982) Cardiovascular adjustments to laboratory diving in beavers and nutria. *American Journal of Physiology* **11**: 434–440.

Miller, R. F., Harington, C. R. and Welch, R. (2000) A giant beaver (*Castoroides ohioensis* Foster) fossil from New Brunswick, Canada. *Atlantic Geology* **36**: 1–5.

Nolet, B. A. and Rosell, F. (1994) Territoriality and time budgets in beavers during sequential settlement. *Canadian Journal of Zoology* **72**: 1227–1237.

Nolet, B. A. and Rosell, F. (1998) Comeback of the beaver *Castor fiber*: An overview of old and new conservation problems. *Biological Conservation* **83**: 165–173.

Nolet, B. A., Van Der Veer, P. J., Evers, E. G. J. and Ottenheim, M. M. (1995) A linear programming model of diet choice of free-living beavers. *Netherlands Journal of Zoology* **45**: 315–337.

Novak, M. (1987) *Beaver*. In M. Novak, J. A. Baker, M.E. Obbard and B. Malloch (Eds.) *Wild Furbearer Management and Conservation in North America*, 283–312. Ontario Trappers Association and Ontario Ministry of Natural Resources, Toronto.

Parker, H., Nummi, P., Hartman, G. and Rosell, F. (2012) Invasive North American beaver *Castor canadensis* in Eurasia: A review of potential consequences and plan for eradication. *Wildlife Biology* **18**: 354–365.

Pilleri, G. (1986) *Investigation of Beavers Vol V*. Brain Anatomy Institute. Berne.

Pope, L. (2013) *Beavers have been busy sequestering carbon!* New Scientist July 17.

Rosell, F. and Sun, L. (1999) Use of anal gland secretion to distinguish the two beaver species *Castor canadensis* and *C. fiber*. *Wildlife Biology* **5**: 119–123.

Rosell, F., Bozsér, O., Collen, P. and Parker, H. (2005) Ecological impact of beavers *Castor fiber* and *Castor canadensis* and their ability to modify ecosystems. *Mammal Review* **35**: 248–276.

Sharpe, F. and Rosell, F. (2003) Time budgets and sex differences in the Eurasian beaver. *Animal Behaviour* **66**: 1059–1067.

Thomsen, L. R., Campbell, R. D. and Rosell, F. (2007) Tool-use in a display behaviour by Eurasian beavers (*Castor fiber*). *Animal Cognition* **10**: 477–482.

Vispo, C. and Hume, I. D. (1995) The digestive tract and digestive function in the North American porcupine and beaver. *Canadian Journal of Zoology* **73**: 967–974.

Willby, N. J., Casas Mulet, R. and Perfect, C. (2011) *The Scottish Beaver Trial: Monitoring and further baseline survey of the aquatic and semi-aquatic macrophytes of the lochs 2009*. Scottish Natural Heritage Commissioned Report No. 455.

Wilsson, L. (1971) Observations and experiments on the ethology of the European Beaver (*Castor fiber* L.). *Viltrevy* **8**: 115–166.

About Pelagic Publishing

We publish books for scientists, conservationists, ecologists and wildlife enthusiasts – anyone with a passion for understanding and exploring the natural world.

You may also be interested in:

Badger Behaviour, Conservation & Rehabilitation - 70 Years of Getting to Know Badgers, George Pearce, 2011

British Bat Calls: A Guide to Species Identification, Jon Russ, 2012

Social Calls of the Bats of Britain and Ireland, Neil Middleton, Andrew Froud and Keith French, 2014

Urban Peregrines, Ed Drewitt, 2014

Amphibians and reptiles, Trevor Beebee, 2013

Barn Owl Conservation Handbook: A comprehensive guide for ecologists, surveyors, land managers and ornithologists, The Barn Owl Trust, 2012

Shrewdunnit: The Nature Files, Conor Mark Jameson, 2014

www.pelagicpublishing.com